REFLECTIONS ON THE

Nature of God

Reflections on the Nature of God
Copyright © 2004 Lionheart Books, Ltd.

Published by Templeton Foundation Press
Five Radnor Corporate Center, Suite 120
Radnor, Pennsylvania 19087

www.templetonpress.org

Library of Congress Cataloging-in-Publication Data
Reflections on the nature of God / edited by
Michael Reagan; introduction by Martin E. Marty.
p. cm.
ISBN 1-932031-69-3

(pbk. : alk. paper) 1. God. I. Reagan, Michael, 1945-

BL205.R44 2004
211—dc22
2004012422

First edition

Printed in China

04 05 06 07 08 09   10 9 8 7 6 5 4 3 2 1

Produced by Lionheart Books, Ltd.,
5105 Peachtree Industrial Boulevard,
Atlanta, Georgia 30341

Design: Carley Wilson Brown & Michael Reagan
Research: Lisa Reagan and Deborah Murphy

*Rings of Saturn—Images taken during the Cassini spacecraft's orbital
insertion on June 30, 2004 show the best view of Saturn's rings in the
ultraviolet spectrum.*

REFLECTIONS ON THE
Nature of God

Introduction by Martin E. Marty

Edited by Michael Reagan

TEMPLETON FOUNDATION PRESS
PHILADELPHIA AND LONDON

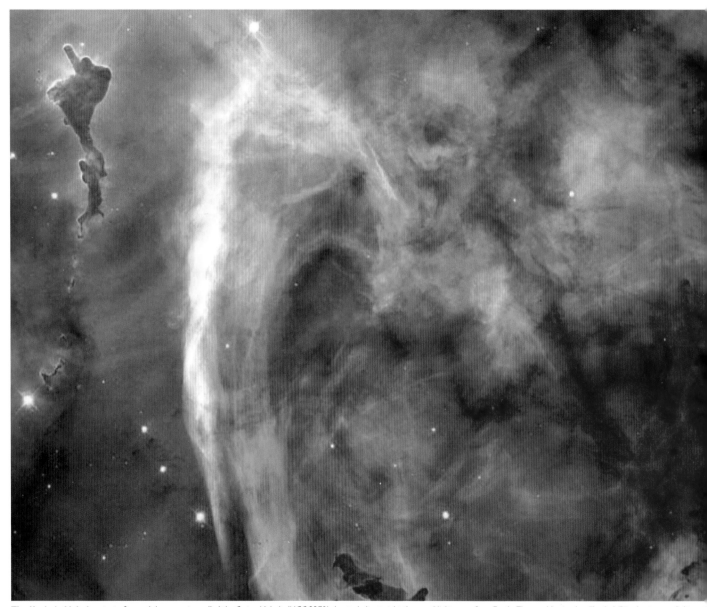

The Keyhole Nebula—part of a much larger region called the Carina Nebula (NGC 3372), located about eight thousand light years from Earth. The roughly circular "keyhole" is about seven light years across and is surrounded by silhouetted clouds of cold molecules, bright filaments of hot gas, and dust. The two pillarlike structures to its left are dust clouds that may eventually give birth to star clusters.

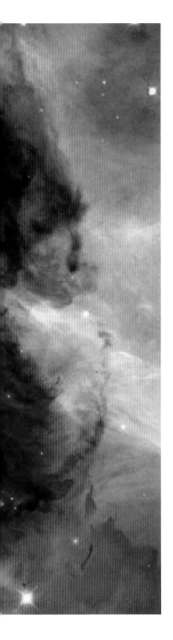

PREFACE

Several years ago I edited a book entitled *The Hand of God: Thoughts and Images Reflecting the Spirit of the Universe,* which included new images of space and the cosmos. In that book, we examined the basic question, "Is there a God?" The overwhelming impression I had after looking at the new images generated from the Hubble telescope was one of awe and wonder, and the feeling that we are not alone. This gave me comfort. In fact, the more we look at this evidence of the basic structure of our universe, the more it appears to scientists and others that there is indeed some hand that guided us into existence. So the next question to explore is, "What is the nature of God and how is this nature revealed—in all senses of the meaning of nature?"

I am still awed by the images from deep space; there is a sense of majesty that is profound, almost beyond comprehension. I sense the presence of God, but it is in the details that I find more comprehensible information about the nature of God. How is it that this obscure planet on the edge of a galaxy in a universe of billions of galaxies has such an amazing abundance of life? The diversity we see here on earth is as astounding as the stars themselves. The interconnectedness of life and its many forms is perhaps the most amazing part of this creation.

As Martin Marty points out in his introduction, it is easy to conclude that one would be either arrogant, naïve, or just foolish to conceive of a book about the nature of God. We would all agree that creating a finite representation of the infinite quality of the Creator of our vast universe is impossible. The intention of this book, therefore, is simply to ask the question, "What is the nature of God?" and to give readers a few moments to reflect on what it means to be a part of this universe. The purpose is also to remind us that there is wonder all around us, and that sense of wonder is the greatest insight of all.

MICHAEL REAGAN

INTRODUCTION

Viewers of the photographs in *Reflections on the Nature of God*, if they pause to reflect as they turn the pages, are likely to be awestruck. Whether the camera reaches to the most distant known places in the universe or focuses up close on small strange creatures, they will have good reason to be dazzled into silence as they view its images. Whether such viewers consider the pictures to reveal much of anything about the nature of God will depend in no small measure on who they are, what they bring to the viewing, and what they want to take from their experience. The well-chosen reflections on God's nature, the little paragraphs accompanying the pictures, will also stimulate wildly diverse reactions. That is how it should be.

Skeptics may think that any venture called *Reflections on the Nature of God* has to result from either naïveté or arrogance on the part of the editor and photographer. I am confident that Michael Reagan would welcome these skeptics at his side and hope to start a conversation about their responses—a conversation that might open with talk about naïveté and arrogance.

Naïveté, first. In one of the oldest but still always illuminating stories from the world of kindergartners we hear of a five-year-old who is engrossed in an art project. We picture her lying on the floor, her hair falling to the sides of her face, perhaps her tongue curled out of one side of her mouth. "What are you drawing?" asks the mother. "I'm painting a picture of God!" No, smiles her mother, "No one knows what God looks like!" To which the daughter, not losing confidence, comes back, "Well, they soon will!" Let the little girl be a reminder that attempts to address or portray the nature of God are by definition impossible, and that this book acknowledges that truth.

As for the other charge of the skeptic, that any such endeavor as this is the product of arrogance, let us recall the story of Eunomius of Cyzicus. This third-century scholar and leader boasted that he knew God better than God did. Why? Because God was busy being God, while his servant Eunomius had the time and the calling to contemplate and study God, and therefore to write intelligently about God. Eunomius is not the patron saint of

Reflections on the Nature of God, and is called upon to do no more than serve as a reminder that undertakings such as these have boundaries and limits.

With naïveté and arrogance out of the way, there are good reasons for readers to let this book work its effects. Ministering humbly to a deep need in human nature, it belongs to an enduring tradition out of which comes guidance for mortals who are at various stages along their way to self-understanding. Self-understanding always involves understanding of "the other." This "other" can be a family member or friend, a teacher or a poet, an alien or an enemy. In most cases, however, across the ages and in the various cultures, the most important "other" gets capitalized. This Other, however conceived, in its deepest depth and furthest reach, is often called God by many who claim or hope that the search for God contributes to human understanding, identity, and their own spiritual quests.

With this book as a guide, let us briefly consider where the search for God often leads, as well as the inevitable difficulties the journey encounters.

DETAIL

The early pages of the book result from camera probes connected with space exploration, thanks to recent advances in telescopy. They are not here to confirm scientific data concerning the measurement of distances. If they did, the mathematics would add little, since the spaces are so vast that those who contemplate them would quickly lose perspective. Still, despite the universe's incomprehensible dimensions, instinct tells small humans in obscure places such as the earth as it revolves within the solar system, that they should reckon with size and distances. How can they undertake this reckoning?

Often in books devoted to great art and artists, the authors and editors set out to inform readers about some feature of a particular work by showing an enlarged close-up of one part of it, and then labeling it "Detail." Thus when a complex scene such as a Brueghel painting of villagers going about their work or play in Holland benumbs the viewer, or wearies her so that she is tempted to turn the page unmindful of its features, we expect to see on a second page a glimpse of a tiny space in the larger picture—perhaps a depiction of an eye or a belt-buckle, labeled "Detail." So it is with a large and distant, universe-encompassing God. A *New Yorker* cartoon some years ago showed an utterly banal scene, a forgettable glimpse of a man heading for work, briefcase in hand, as he waited on a corner, no doubt for a bus. The artist captioned it: "The Milky Way: Detail," which, of course, it was. So would be a glimpse of a ring of Saturn, a picture of an atom, or a discarded toothbrush in a garbage dump: "The Milky Way: Detail." Would they, taken together and multiplied by billions of billions of billions of additional snatches of reality from these photographs add up to "God"? It would be hard to picture any serious person contending that they would.

While words and pictures cannot "prove" the existence of God, they can evoke responses that resonate in the souls of seekers. Through the ages philosophers, theologians, poets, and ordinary humans, have taken the measure of their own lives by standing under a starry sky. They usually remark that the experience both humbles and enobles them. The view is humbling, since, even without precise mathematical accounts of distances, the human is quickly made aware that something bigger than his or her own little life is occurring. And it is enobling, since the one who does the viewing, imagining, and speculating, is equipped to do all of those things in unique ways. Similarly, poets look at the small and the fleeting: "God's" creation, as in Percy Bysshe Shelley's "To a Skylark," or marks of human creativity as in Keats' "Ode to a Grecian Urn." Enhancements of the human imagination can provoke readiness for the experience of God, an experience which may or may not follow. Still, anticipating such, or wondering, or letting poetic and philosophic thoughts erupt is, according to all reports, intrinsically valuable. We hope this book will elicit "intrinsically valuable" responses—and more!

 ## DARKNESS

A God lost in and beyond the details would also be lost in the features of the universe that cannot be disclosed by the photographer and are barely conceivable to the philosopher. Much of the surface area of the pictures in the forepart of this work is blotted out in blackness. That blackness and what it represents, about which we know less than did the ancients, because we know more about it scientifically, terrified earlier contemplators. In his *The Four Quartets* T. S. Eliot wrote of them and us:

> *O dark dark dark. They all go into the dark,*
> *The vacant interstellar spaces, the vacant into the vacant, . . .*
>
> *I said to my soul, be still, and let the dark come upon you*
> *Which shall be the darkness of God.*

Where does reflection on darkness get us in a search for the nature of God? When the prophet Isaiah declares in the name of the biblical God, "I form light and create darkness," the thoughtful but

Hubble Ultra Deep Field—*contains stars that may have emerged during the so-called pre-Big Bang "dark ages" that preceded the well-lit universe we know today. The section shown here lies below the*

constellation Orion and contains an estimated 10,000 galaxies which existed 13 billion years ago, when the universe was only five percent of its present age.

believing reader is not given freedom to equate God with either light or darkness, both of which are seen to be God's creation.

Scientists today speak of "dark matter" as constituting much of the universe. Some astrophysicist devotees of the Big Bang theory of the origin of the universe hypothesize, or contend, that in that origin, matter and dark matter were in symmetry, but matter was perhaps one trillionth heavier than its opposite. For that reason, there is something as opposed to nothing, or at least as opposed to impenetrable and undefinable dark matter. Whether such mathematics and such a hypothesis are widely shared, or whether or not they stand up well, language at least points to scientific efforts and limits. Scientists who do such speculating tend to be modest and always have to be ready to revise their theories. The great physician-writer Dr. Lewis Thomas famously said that "the greatest of all the accomplishments of twentieth-century science has been the discovery of human ignorance." Dr. Thomas, a great scientist, did not write a line like that in an effort to help abort the scientific quest but rather to point to its side-effects. One of these, in our context, has to be that science, properly conceived, points to the human ignorance that prevailed when in the past people dealt with God spatially. They confined God within much too small dimensions as they attempted to prove His existence by suggesting that God resided just beyond the light we see and know.

LIGHT

Light, needed for reflection, naturally works better than darkness in the imaginations of those who would reflect on the nature of God. So the quest for the nature of God in this book and elsewhere figuratively pursues the realm of light. In that sphere humans become aware of their limits and have reason to glory in their new awarenesses. In some faiths, God is Light and all light issues from God.

My friend David Schramm, a noted astrophysicist, one day twitted me—as if I were a fundamentalist "Creationist"—twitted me with the question, "Marty, what is it like to be confined in a universe created in six days six thousand years ago, ever after which nothing happened?" My answer: "You have it wrong. In the biblical tradition we believe in *creatio continua*, which implies ongoing creation. This means that however things got started, the light spoken of in the Genesis story still shines and gives us all the energy and illumination off which we and all things live and will live."

Then it was my turn: "David, what is it like to be confined in a universe that banged itself into existence in a nanosecond many billions of years ago, after which nothing happened?" He responded: "You have it wrong. Ever after the Big Bang, the light there generated still shines and gives us all the energy and illumination off which we and all things live and will live." (I doubt whether the exchange was as symmetrical as my account suggests, but this was the gist in both cases.) An eavesdropping colleague added his word: "As I hear it, we have here parallel but not mutually contradictory assertions, using languages that have different intents in both cases."

✠ MINUTENESS

The nature of God as a large Being often gets matched by the notion that God is also ubiquitous, everywhere—as pantheists believe—or present in the smallest, nearest, parts of creation as well as the largest and most distant. At the turn of our new century some schools of scientists devoted themselves to what is smallest. Best known among them are superstring theorists, who ask

The Ant Nebula (Mz3)—*resembling the head and thorax of an ant, is a dying star located between 3000 and 6000 light years from Earth. As seen by the Hubble Space Telescope, the fiery lobes protruding from Mz3 form the "ant's" body.*

and try to answer the question: What are the smallest elements of nature, the finally indivisible components? Are they quarks, electrons, particles? Or, as superstring theory would have it, while these other components do their work as dots, "smallest" now refers to superstrings, or particles made up of tiny filaments of energy. Brian Greene, one of the proponents of this controversial theory, says that these string-shaped filaments are "some hundred billion billion times smaller than a single atomic

spatial references. Ever more scientists, mystics forever, and many religious scholars caution against a too ready resort to dimensional schemes and measurement. In a book on Italian church architecture published after World War II, the authors posed on facing pages two cathedrals of roughly equal size. One towered above the huts and cabins of European villages in a medieval cathedral city. The other was St. Patrick's in New York, viewed from an airplane, the church nestled among and almost obscured by the skyscrapers of and surrounding Rockefeller Center. The caption, roughly translated, said that in the "volume-try" of the modern city, the religious presence had to make its effect not "dimensionally" but "substantively."

Whatever else that meant, it was a chastening reminder that what one era or people in one place consider to be large, those in another no longer will. So with the age-less effort to see God as so large as to appear beyond current reach, behind the curtain of the heavens, or as small as tiny filaments, hidden in the shrouds of human instruments and intelligence. As the book on architecture said of buildings, now we would say that in speaking of God there has to be a substantive reference, some "content." Talk about God or viewing pictures suggesting God usually implies some sort of content, something more than hypotheses. How does one best do this suggesting, some critics ask. The pictures here cannot do the work alone and, at their best, they represent only a beginning. However, what is here comes as an invitation to ponder, to reflect on God as Creator. Most people who claim to have experienced God or who have philosophized about God would say the Creator did and does more than merely create and take the rest of the day or the rest of all time off. Still, reflections about the nature of God as Creator usually come first.

nucleus." They vibrate in different patterns. At present one has to accept such a theory partly on faith, since it is not testable: No one possesses and we may never possess an instrument to discern and test their presence. Yet, if they exist, they serve well as a reminder that, in the pictures of nearby creatures that pose in this book, we are not going to locate God in the closest of close-ups.

SUBSTANCE

Everything about which I have written so far locates the search for the nature of God in space, or uses

If the pictures serve to evoke in the imagination anything that encourages reflection on the nature of God, they no doubt do so most in the minds of those who have already heard particular stories from those who listened to the words of myth-announcers, mystics, prophets, and revealers—i.e., stories that embue God with substance. Apart from those words, what appears in this book would be nothing more than pixels on the page. Some would be light and some dark, some fuchsia and some aquamarine, some defined and some blurry, some as close as a reptile's eye and some as distant as celestial objects billions of light years away; some enduring and others as ephemeral as one droplet in a pictured wave's crest; some tender and some evoking the everyone-eating-everyone-else implication that the nature of God includes the creation or permission of cruelty. These pages almost inevitably address what is in the mind of viewers and readers as a result of their prior teaching and preexistent curiosity.

▨ VARIETY

Reflections on the Nature of God necessarily includes the diversity of species. If God is one, as so many religions profess, God certainly uses divine imagination in creating or permitting or seeing the development of species. The pages in this book that are devoted to creatures large and small can only suggest how varied living things are. Scientist J. B. S. Haldane, told that there were 300,000 species of beetles alone, once commented that God has "an inordinate fondness for beetles." If others connect the idea of God with the number of species, they soon stare into an abyss of yet-to-be-discovered species and hard-to-count discovered ones. The United Nations Environment Program's Global Biodiversity Assessment counts about 1,750,000 described species, many of which are so complex and alluring that they seem indescribable. Scientists who work in the middle

The mantis shrimp is not really a shrimp, but a predatory crustacean called a stomatopod. Some varieties, called smashers, are capable of striking their claws together to produce a blow equivalent to the shot of a .22 caliber bullet. Others spear their prey, striking quickly and powerfully enough to sever a human finger.

Migrating wildebeest herd and giraffe on a savanna in Africa.

range of estimates speak of five to twenty million species, most of them unknown as of now.

Students of nature, who remember that the human species is but one among so many, may comment that the human is therefore quite insignificant in the scheme of things. Just as readily, others agree with philosopher Blaise Pascal that, for all the limits, the human being is a "thinking being," the only one of which we know that reflects on the nature of God. The thinking beings have to do much counting, for diversity is very much a part of human experience and perception. The numbers of religions, whose members do the reflecting on the nature of God, runs into the many tens of thousand. Each of them

offers at least a subtle difference, sometimes a drastic difference, from all others. It is not surprising that so often both those who are formally religious and those who are not are often united chiefly by their experience of the kind of awe this book's pictures, of scenes celestial and creaturely, evokes.

LANGUAGE

Little wonder that humans so regularly try to communicate this sense of awe and what they experience as divine by resorting to language. While reflection on God can prosper by means of response to examples and genres, it usually issues in the form of language, of words that give voice to ideas. Here is where literature, as represented by the quotations included on these pages, comes in. Yet readers will remain dissatisfied if they desire what are linguistic equivalents of the little girl's announcement that, after viewing her work, others would know what God looks like. No sure-of-himself Eunomius of Cyzicus is represented, nor are absolutists, dogmatists, heresy-hunters, or over-definers. That is not to say that the writers here lack conviction, but they make their statements in provocative ways that are designed to stimulate thought and conversation, not to end them abruptly by silencing the questioner or seeker.

NOT

The approach Michael Reagan and, in this introduction, I have taken has some background in what is technically the "apophatic" approach. In other words, so far on my pages and through most of his, there is pictorial or verbal comment on the nature of God by saying what God is not. God is not the dark matter or the explosions of light, not the distances or the nearnesses, the telescopic probes or the close-ups of the tiny forms of life. God may be "in" them, and to the pantheist, they are "in" God. But there are so many other kinds of theism than pan-: monotheism, polytheism, panentheism, and more, and readers devoted to any of these are to be as welcome in this probe as are pantheists.

The "apophatic" way is a method of denying something in order to stimulate responses in the form of heightened awareness. The apophatic advocates say what God is not in order to reinforce the idea of divine transcendence. Some of them simply say that all words and pictures taken together cannot do justice to the reality. God is beyond human description, yet recognizing the limits of all attempts is a way of stimulating a sense of awe, of divine glory, wonder in the face of the Holy, the Sacred language, the Other.

Obviously, *Reflections on the Nature of God* intends more than a discussion of what God is not, and it is in place now to ask what its intentions are. I see two:

WONDER

First, to stimulate wonder. *Reflections on the Nature of God* sets out to start a conversation. Much talk about the nature of God among people of no religion or many religions takes the form of argument. Argument has its place in legislatures, science classrooms, and among theologians who agree on the authority of a particular text, as when Muslims exegete the Qur'an. How can anyone intelligently argue, however, when there is no agreement on what the disagreement is? That conundrum has to prevail in this case, since there is no way for humans to speak from God's point of view and thus settle once and for all which religion, which means which way of speaking of the nature of God, is the true one.

Conversation, on the other hand, begins with the question: "What do you see in these pictures that might lead you to reflect on the nature of God?" "What do these quotations do to push you to fresh thought on God's nature?"

Conversations thus begun are never ending. They have the character of play. No one says, "I surely won that

conversation!" Instead, by opening a conversation, speaker A invites speaker B to respond, and a kind of game—sometimes a deadly serious game and sometimes a merely lively one—ensues. Because the participants never know quite where the conversation will go, they remain alert, inquisitive, perhaps ready to change. And many do change.

Such conversation, prompted by these pictures and the accompanying quotes, should promote a sense of wonder. Those who present this book to you obviously think that such a sense is, or at least can be, a human good. Forty years ago, when desire and action among many to control the environment and master the graspable universe was in special vogue, some theologians (i.e., speakers about God) questioned the value of wonder. One of these thinkers remembered how giants in philosophy like Immanuel Kant or in mathematics and theology like Blaise Pascal stood under the starry sky and were moved to think about transcendence, about otherness, anything which is beyond control and human mastery, in other words: God.

Carved primarily by the Colorado River over the past six million years, the strata along the walls of the Grand Canyon form a geological record dating back to the Precambrian era, 1.7 billion years ago.

That same thinker reported on an experience he had in the early days of Echo and other satellites that had been lofted by technological wizards. He was in his back yard with his son, who asked, "Which ones did we put up there?" Gone was wonder, unless in the form of wonder over rocket engines and instruments. Such a reduction of wonder did not end the search for further scientific control and mastery. Impulses toward them remained and, demonstrably if ambiguously, they served human good. However, that search did not address many of the yearnings, desires, and needs of the human heart and soul. The sense of wonder by itself will not satisfy all of these. But it will help render humans humble and thus ready to extend their curiosity to matters of soul and spirit, to science and arts, to acts of mercy and justice.

◈ RESPONSE

For that reason, while seeking to minister to the impulse to wonder, *Reflections on the Nature of God* has as its intent the evocation of response. All of the pictures in this book deal with the natural, not the manufactured world. No action now conceivable on earth can affect the heavens, galaxies away. However, those who wonder in the face of the beauty and terror of space or the complexity and elusiveness of the plant and animal world, have new motivation to respond with care for what is in range, meaning the earth and all that is in it.

Nothing herein is designed to diminish humans, though it may play a small part in helping them relocate themselves in the natural universe. Sometimes that which is large is supposed to stun that which is small. A very large wife in a cartoon, standing next to a diminutive, bald, bespectacled husband at the edge of the Grand Canyon, says "Makes you feel very insignificant, doesn't it?" Those who respond in wonder reach beyond the boundaries of the moment in time, the speck in place, and find new significance.

Philosopher Martin Buber once said that God is addressed, not expressed. Is the nature of God pointed to on these pages such that one would address, pray to, experience, feel the weight of, or delight in the release

from, this God? People of various faiths who speak of their connection to God not as to "It," but, as Buber has it, as to a "Thou," will reflect on this question in different ways. For instance, as a Christian I will be inclined to testify to a line Paul wrote in a letter to the Colossians, that "all things,"—which would mean all things in *Reflections on the Nature of God* —"have been created through [Christ] and for him . . . and in him all things hold together." (Colossians 1:17) The apostolic writer understood that his words would have little meaning in the eyes of those who did not believe in Christ.

Something like that is the case with the particular stories about the object of faith in all the religions. None of these stories are confirmed in the wonderful and wondrous pictures in this book. Moreover, no believer is likely to be converted by this book to a different religion, another way of speaking of God. Yet if the readers of this book are moved to wonder, their response should at least include an enlargement of their imagination, a stimulus to their creativity, a motivation for them to reencounter the texts and stories of the faiths and the sciences, and, one hopes, a readiness to be more open than before to the presence of the neighbor or the stranger who also seeks and ponders "the nature of God" in a world of immense diversity and profound longings that will not be denied.

MARTIN E. MARTY
Fairfax M. Cone Distinguished Service Professor Emeritus
The University of Chicago

And out of nothing, and into nothing, God, by a free decision, set up the spontaneous production of particles, in newborn space and time, producing a silent, seething sphere, infinitesimally small and unimaginably hot. There was onset and evolution, the first stage of creation.

During a tiny fraction of a second, an expansion took place, and the perfect symmetry of the forces was broken up, step by step, as the temperature dropped, to produce the forces of nature we know today.

God's well-tuned laws made innumerable particles, of every requisite kind, in a steadily expanding chaotic cooling sphere. And the universe cooled for nearly a million years, until electrons could stay joined to nuclei to form familiar atoms. There was onset and evolution, the second stage of creation.

With atoms and moleules as building blocks, the attracting force of gravity took over, and after about a thousand million years, God saw the first stars and galaxies forming in an expanding cosmic universe. There was onset and evolution, the third stage of creation.

Individual stars contracted under gravity and became hot enough for nuclear fusion to produce chemical elements not seen before, until, after about ten thousand million years, stars were exhausted by their radiance, and God saw them begin to die, some dramatically, by exploding as supernovas, releasing all the known chemical elements. There was onset and evolution, the fourth stage of creation.

And God saw that it was very good, for now all the ingredients were available, and gravity formed a second generation of stars, some accompanied by planets and satellites, including the Sun, Earth, and later, the Moon, in our galaxy of the Milky Way. There was onset and evolution, the fifth stage of creation.

Bathed in alternate daylight and darkness, during the next thousand million years or so, conditions on Earth became favorable for the eventual generation of life. There was onset and evolution, the sixth stage of creation.

During these last three thousand million years, life has evolved as God intended, and through numerous cycles of birth, survival, procreation, and death, species have multiplied and progressed, plants and animals of every kind, and some have become extinct, until, a mere three hundred thousand years ago, there arrived, in the likeness of God, *Homo Sapiens*, intelligent humans, with freedom to choose, living together in community, knowing good and evil, pleasure and pain, aware of honor due to their dominion, and acquainted with death. There was onset and evolution, the seventh stage of creation.

Eta Carinae —*During its massive outburst 150 years ago, Eta Carinae became one of the brightest stars in the southern sky. More than 8,000 light years away, it radiates about five million times more power than our Sun, and can be seen in the Eta Carinae nebula, NGC3372.*

And the universe entered the Age of Humanity.

TED BURGE
A Creation Story for Our Times

We are stardust,
We are golden,
We are billion-year-old carbon,
And we've got to get ourselves
Back to the garden.

JONI MITCHELL

Detail of M42—the Orion Nebula, discovered in 1610 by Nicholas-Claude Fabris de Peiresc. Covering four times the area of the full Moon, M42 is the brightest and most photographed nebula in the night sky.

Star Formation in RCW49— (right) One of the most prolific birthing grounds in our Milky Way galaxy. Located 13,700 light years away in the southern constellation Centaurus, RCW 49 is a dark and dusty stellar nursery that houses more than 2,200 stars.

This image taken by the new Spitzer Space Telescope's infrared array camera highlights the nebula's older stars (blue stars in center pocket), its gas filaments (green) and dusty tendrils (pink). Speckled throughout the murky clouds are more than 300 never-before-seen newborn stars.

If the mass of the neutrino were not precisely tuned there would be no Earth-like planets and hence no life as we know it. We are indeed children of stardust, stardust powered on its journey through the cosmos on the wind of neutrinos.

 SHARON BEGLEY
The Hand of God

The Pelican Nebula (W80)—*located in the constellation Cygnus, it is about 1,800 light years away from earth.*

Messier 64 *has a spectacular dark band of dust in front of the galaxy's bright nucleus. At 17 billion light years away from Earth, M64 is nicknamed "The Black Eye" or "The Evil Eye." The small blue dots, left, are stars that have just formed.*

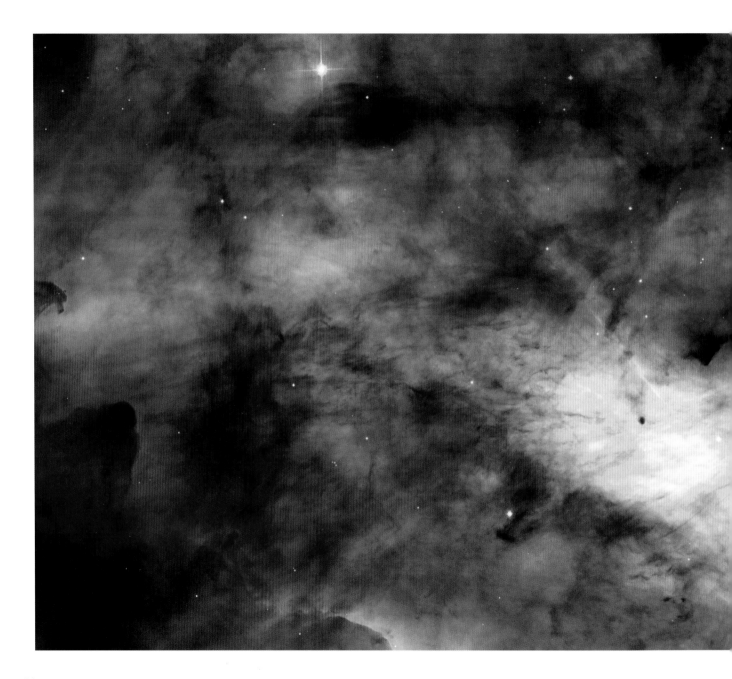

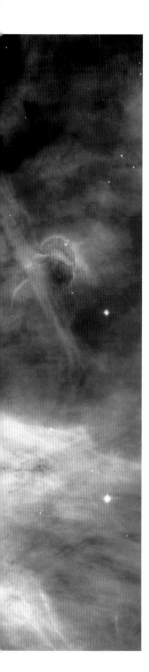

Physical processes come in two varieties—lawful and random. Traditionally, scientists assumed that the origin of life was a chemical fluke of stupendous improbability, a quirk of fate unique in the entire cosmos. If so, then we are alone in an otherwise sterile universe, and the existence of life on Earth, in all its exuberant glory, is just a meaningless accident. On the other hand a growing number of scientists suspect that life is written into the fundamental laws of the universe, so that it is almost bound to arise wherever earthlike conditions prevail. If they are right—if life is part of the basic fabric of reality—then we human beings are living representations of a breathtakingly ingenious cosmic scheme, a set of laws that is able to coax life from nonlife and mind from unthinking matter. How much more impressive is such a magnificent set of physical principles—which bear all the hallmarks of design—than the sporadic intervention of a Deity who simply conjures these marvels into existence.

PAUL DAVIES

Closeup of the center of the Omega Nebula, also called the Swan Nebula, M17. The region shown is about 3,500 times wider than our solar system, and resides 5,500 light years away in Sagittarius.

Do we want to contemplate his power? We see it in the immensity of the Creation. Do we want to contemplate his wisdom? We see it in the unchangeable order by which the incomprehensible WHOLE is governed. Do we want to contemplate his munificence? We see it in the abundance with which he fills the earth. Do we want to contemplate his mercy? We see it in his not withholding that abundance even from the unthankful. In fine, do we want to know what GOD is? Search not written or printed books, but the Scripture called the "Creation". . . .

. . . God is the power of first cause, nature is the law, and matter is the subject acted upon. . . .

. . . I have said in the course of this discourse, that the study of natural philosophy is a divine study, because it is the study of the works of God in the creation. If we consider theology upon this ground, what an extensive field of improvement in things both divine and human opens itself before us! All the principles of science are of divine origin. It was not man that invented the principles on which astronomy, and every branch of mathematics, are founded and studied. It was not man that gave properties to the circle and the triangle. Those principles are eternal and immutable. We see in them the unchangeable nature of the Divinity. We see in them immortality, an immortality existing after the material figures that express those properties are dissolved in dust.

THOMAS PAINE

The great mystery is not that we should have been thrown down here at random between the profusion of matter and that of the stars; it is that from our very prison we should draw, from our own selves, images powerful enough to deny our nothingness.

 ANDRÉ MALRAUX

It does nothing, yet it has fashioned the universe. Sustaining the entire universe, it does nothing at all. All substances are non-different from it, yet it is not a substance; though it is non-substantial it pervades all substances. The cosmos is its body, yet it has no body. . . that infinite consciousness is and is not. It is even what it is not. All these statements about what is and what is not are based on logic, and the infinite consciousness goes beyond truth, beyond logic.

 VASISHTHA'S YOGA

AM 0644-741—*A member of the "ring galaxies," AM 0644-741, lies some 300 million light years from earth. The blue ring surrounding it, made up of rampant star formation, is 150,000 light years in diameter, larger than the entire Milky Way.*

NGC7319, NGC7318B and NGC 7318A—*A close-up of three of the five compact galaxies (NGC7319, NGC7318B and NGC 7318A) that make up the galactic group, Stephans Quintet, each cluster harboring millions of stars. The clusters are relatively young between about 2 million to 1 billion years old and reside in the constellation Pegasus, 270 million light years from Earth.*

N44C—*(next page) a region of ionized hydrogen gas surrounding an association of young stars in the Large Magellanic Cloud, a small companion galaxy to the Milky Way. N44C is about 160,000 light years from Earth and roughly 126 light years across. The glowing filaments that resemble tendrils of hair are emitted by hot young stars in a nearby galaxy.*

We shall not cease from exploring,
And the end of our exploring
Will be to arrive where we started
And know the place for the first time.

T. S. ELIOT

NGC 3370—*Similar in size to our own Milky Way, spiral galaxy NGC 3370 lies 98 million light years away from Earth, in the constellation Leo. This composite photograph, the result of 24 hours worth of exposures, is one of the deepest images Hubble has ever made.*

An artist's computer-generated version of a recently discovered giant string of galaxies, approximately 300 million light years long, 50 million light years wide, and lying 10,800 million light years away toward the constellation Grus. We see it here as it appeared 10.8 billion years ago.

There are really two sorts of gods that have been around for some time. We might crudely say that one of these, the personal god, is the god concerned with human behaviour, morality, preservation of life after death. This is a guardian-angel type of god. I would love to believe there is such a being, but I find it very difficult to do so. Then we've got the other sort of god; much more remote, much more powerful, in a sense much more awesome. This is God the great architect of the cosmos, an abstract, timeless being that is somehow responsible for the overall organisation and structure and lawfulness of the universe. . . .

. . . modern physics, and indeed one strand of the Christian tradition going back to Augustine, insists that time itself came into existence with the big bang; the universe began with time and not in time. That point is absolutely fundamental. We are not talking about some superbeing or force or power that was there before the universe, "before" in the temporal sense. I reject the notion of a being who floats there for all eternity, then presses a button, so that, bang!, the universe appears, and then sits back to watch the action, or maybe interferes with it from time to time. So I am not talking about something that was there before. I am talking about something which is timeless something outside of time, indeed space, altogether.

 PAUL DAVIES

The Great Orion Nebula—(M42, NGC 1976) and NGC 1977 (on the left) are located in the "Sword" part of the constellation of Orion, just below the easternmost of the three stars that comprise Orion's belt. At 1,500 light years away, this turbulent, star-forming region of dust and gases was first cataloged in 1764.

When we consider the first seconds of the Big Bang that created the universe, it is an astonishing reflection that at this critical early moment in the history of the universe, all of the hydrogen would have turned into helium if the force of attraction between protons—that is, the nuclei of the hydrogen atoms—had been only a few percent stronger. . . . No galaxies, no stars, no life would have emerged. It would have been a universe forever unknowable by living creatures. A remarkable and intimate relationship between man, the fundamental constants of nature and the initial moments of space and time seems to be an inescapable condition of our existence.

 BERNARD LOVELL

Sheets of debris from a stellar explosion in the Large Magellanic Cloud galaxy have traveled at velocities of more than four million miles per hour to create these delicate filaments. Explosions such as these may have been visible from Earth around 1000 BC.

NGC 6822—*An irregular galaxy, NGC 6822 lies 1.6 million light years from Earth in the constellation Sagittarius. Red stars seen within the body of the galaxy are red supergiants, while bright red stars outside the galaxy are foreground stars in our Milky Way Galaxy.*

To every discerning and illumined heart it is evident that God, the unknowable Essence, the divine Being, is immensely exalted beyond every human attribute, such as corporeal existence, ascent and descent, egress and regress. Far be it from His glory that human tongue should adequately recount His praise, or that human heart comprehend His fathomless mystery. He is and hath ever been veiled in the ancient eternity of His Essence, and will remain in His Reality everlastingly hidden from the sight of men. "No vision taketh in Him, but He taketh in all vision; He is the Subtile, the All-Perceiving." No tie of direct intercourse can possibly bind Him to His creatures. He standeth exalted beyond and above all separation and union, all proximity and remoteness. No sign can indicate His presence or His absence; inasmuch as by a word of His command all that are in heaven and on earth have come to exist, and by His wish, which is the Primal Will itself, all have stepped out of utter nothingness into the realm of being, the world of the visible.

 BAHÁ'U'LLÁH, KITÁB-I-IQÁN

The Clerun-Gowph called the Creator It because the Creator had
no sex of course. . . . (It was a big blue cloud). . .

. . . "But why did It create us!" Simon cried.
"Look at the universe. Obviously, it was made by a scientist, otherwise it
wouldn't be subject to scientific analysis. Our universe, and all the others
It has created, are scientific experiments. It is omniscient. But just to make
things interesting, It, being omnipotent, blanked out parts of Its mind.
Thus, It won't know what's going to happen.
"That's why I think it did not come back after lunch. It erased even the
memory of Its creation, and so It didn't even know It was due back for an
important meeting with me. I heard reports that It was seen rolling around
town acting somewhat confused. It alone knows where It is now, and perhaps not even It knows. Maybe. Anyway, in whatever universe It is, when
this universe collapses into a big ball of fiery energy, It'll probably drop
around and see how things turned out."

KILGORE TROUT (A.K.A. KURT VONNEGUT)
Venus on the Half-Shell

*An artist's rendering of a quasar located in a primeval galaxy.
The iron found in such quasars is believed to have come from
the ashes of the first generation of stars, those that formed
about 200 million years ago, after the Big Bang. Light from
these quasars has traveled for 12.8 thousand million years
before reaching the Hubble Space Telescope.*

The odds against a universe like ours coming out of something like the Big Bang are enormous. I think there are clearly religious implications.

 STEPHEN HAWKING

V838 MonoceratisIn—*this recent view of the star V838 Monoceratis, 20,000 light years away from Earth, the red supergiant in the middle of the image has illuminated the dust and gas, likely ejected in a previous explosion that occurred tens of thousands of years ago.*

Elephant's Trunk Nebula—*Infrared imaging penetrates the dense gas surrounding the ghostly Elephant's Trunk Nebula to reveal embryonic stars and previously unseen young stars. The nebula sits within a larger nebula called IC 1396, in the constellation of Cepheus.*

The Tao has its reality and its signs but is without action or form. You can hand it down but you cannot receive it; you can get it but you cannot see it. It is its own source, its own root. Before heaven and earth existed it was there, firm from ancient times. It gave spirituality to the spirits and to God; it gave birth to heaven and to earth. It exists beyond the highest point, and yet you cannot call it lofty; it exists beneath the limit of the six directions, and yet you cannot call it deep. It was born before heaven and earth, and yet you cannot say it has been there for long; it is earlier than the earliest time, and yet you cannot call it old.

 CHUANG TZU

Where there is freedom, there is the possibility of creation. There is also the possibility of tragedy. The freedom of man to respond to the lure of God is also the freedom not to respond. Freedom is man's opportunity and his tragedy. But the possibilities of value achieved must outweigh the risks of tragedy. There is a tragic element in human life and tragedy in nature. Again the principle of freedom at all levels of creation implies the possibility of chaos instead of order. This is a cost of creation. For God to control the world completely, to take away its freedom and spontaneity, would be to destroy it . . . .

I suggest that God does not proceed as [Bertrand] Russell would like him to, because he is a God of creation, and not a magician. At each stage of the creative process, there are limitations on what can be actualized in the immediate future. The opportunities are limited by what has already been achieved. The existence of man was not a possible immediate step following the origin of the first "living molecule." One stage builds on the previous one to create a continuum. The rivers of creativity do not cut arrows direct to the sea, but wind their way by a long and sure route. As well might we ask why a child does not come into the world as a physically and emotionally mature man! That would be to deny the very nature of creation which is spontaneity of response, purpose, struggle and achievement . . . .

We think of God as revealing himself in special acts at distant intervals. But if God is love, then he reveals himself at all times in all ways and not just in special acts. If all existence is grounded in God, then all existence is a medium of revelation of the nature of God. If man is a vehicle of revelation then all men are, and so is all history and not just the history of one group of people at one particular time.

L. Charles Birch
Nature and God

Instead of picturing God as a medieval monarch on a marble throne, imagine God as the living awareness in the space between the atoms, "empty" space that makes up about 99.99% of the universe. Thinking of God that way gets us past some of the great theological divides of the past. Is God immanent or transcendent, internal or external, composed or compassionate? Like the question of whether the atom is wave or particle, the answer is: yes.

 TOM MAHON

Eye cannot see him, nor
words reveal him;
by the senses, austerity, or
works he is not known.
When the mind is cleansed
by the grace of wisdom,
he is seen by contemplation—
the One without parts.

 MUNDAKA UPANISHAD

Sombrero Galaxy—*Called the Sombrero galaxy because of its resemblance to a Mexican hat, Messier 104 (M104) lies at the southern edge of the rich Virgo cluster of galaxies and is equivalent to 800 billion suns. It measures 50,000 light years across and is located 28 million light years from Earth.*

There may be some problems about how much sense the Universe makes, how rational it is, and how much we can understand it. One thing is certain, and this is that the Universe is fantastic! It is a high level of imaginative power . . . a fantastic effort of creativity.

 FATHER THOMAS BERRY

Juvenile Batfish, Solomon Islands

The Red Spider Nebula NGS 6537—*3000 light years from Earth in the constellation Sagittarius. A stellar wind speeding from a central hot white dwarf (star), at about four to ten million mph, has created the luminous ripples in the gas of the nebula, just as wind would create a disturbance on the surface of a lake on Earth.*

The forms and creeds of religion change, but the sentiment of religion—the wonder and reverence and love we feel in the presence of the inscrutable universe—persists.

 JOHN BURROUGHS

ngels are amazed when they hear that there are men who attribute all things to nature and nothing to the Divine, and who also believe that one's body, into which so many wonders of heaven are gathered, is a product of nature. Still more are they amazed that the rational part of man is believed to be from nature, when, if men will but raise their minds a little, they can see that such things are from the Divine and not from nature, and that nature has been created simply for clothing the spiritual and for presenting it in a corresponding form in the ultimate of order. Such men they liken to owls which see in darkness, but in light see nothing.

EMANUEL SWEDENBORG

Distant galaxies dot the background in this image of disrupted spiral galaxy Arp 188, also called the Tadpole Galaxy. The cosmic tadpole is 420 million light years away, its tail about 280 thousand light years long. The bright blue star clusters in the tail will form smaller satellites of the galaxy which will fall away as the tadpole grows older and "loses" its tail.

White water lily opened to the light.

I am the dust in the sunlight, I am the ball of the sun…
I am the mist of morning, the breath of evening…
I am the spark in the stone, the gleam of gold in the metal…
The rose and the nightingale drunk with its fragrance.
I am the chain of being, the circle of the spheres,
The scale of creation, the rise and the fall.
I am what is and is not…
I am the soul in all.

RUMI

All things by immortal power
Near and far
Hiddenly
To each other linked are.
That thou canst not stir a flower
Without troubling of a star.

FRANCIS THOMPSON

It is easy to understand God as long
as you don't try to explain him.

JOSEPH JOUBERT

The Anti-Atlas Mountains of Morocco formed about 80 million years ago in a collision which destroyed the Tethys Ocean. Layers of the ocean bed folded and crumpled to create the mountains; here, short wave-length infrared bands combine to dramatically highlight the different rock types: limestone, sandstone, gypsum, and granite.

Men are not flattered by being shown that there has been a difference of purpose between the Almighty and them.

ABRAHAM LINCOLN

I simply haven't the nerve to imagine a being, a force, a cause which keeps the planets revolving in their orbits, and then suddenly stops in order to give me a bicycle with three speeds.

QUENTIN CRISP

In this artist's concept, the very first stars in the universe burst into life in a torrential firestorm as brilliant as a fireworks finale. In this case, the finale came long before the emergence of the Earth, the Sun or even the Milky Way.

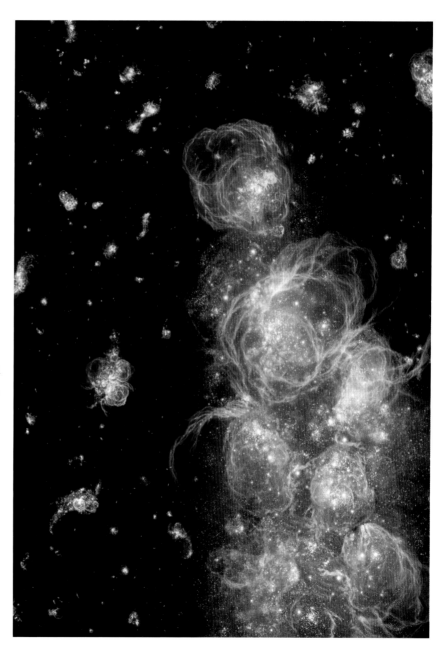

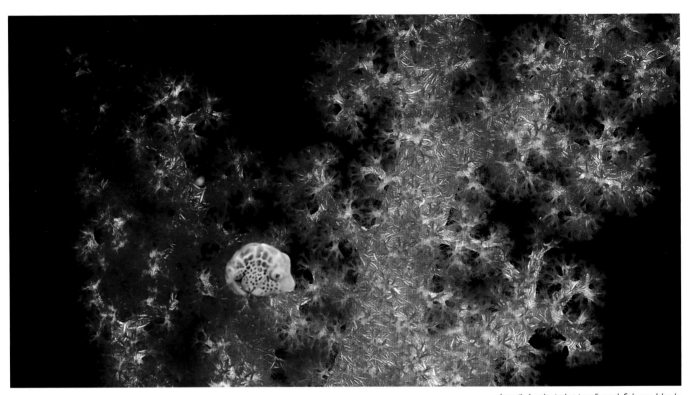

Juvenile Leatherjacket in soft coral, Solomon Islands.

It is the nature of God to reside in mystery—ineluctable,
inexhaustible mystery. We do not need to understand the cabala of
mathematical physics to comprehend the *mysterium tremendum.*
We need only look out the window.

 **CHET RAYMO**
Skeptics and True Believers

Dwarf Galaxy NGC 1569 *is a fertile star factory, as
evidenced by the brilliant blue star clusters seen here. The
sudden onset of star birth began about 25 million years ago,
with the advent of our earliest human ancestors on Earth.*

Divinity is not playful. The universe was not made in jest but in solemn incomprehensible earnest. By a power that is unfathomably secret, and holy, and fleet. There is nothing to be done about it, but ignore it, or see.

ANNIE DILLARD
Pilgrim at Tinker Creek

Hurricane Floyd just off the Florida coast on September 14, 1999. Sustained winds reaching up to 140 mph extended 125 miles from the eye of the storm.

I am the Lord, and there is none else. I form the light and create darkness, I make peace and create evil. I am the Lord, that does all these things.

 ISAIAH 45: 6-7

I hold (without appeal to revelation) that when we take a view of the Universe, in it's [sic] parts general or particular, it is impossible for the human mind not to perceive and feel a conviction of design, consummate skill, and indefinite power in every atom of it's [sic] composition…it is impossible, I say, for the human mind not to believe that there is… a fabricator of all things.

Of the nature of this being [God] we know nothing.

 THOMAS JEFFERSON

NGC 613— *is a barred spiral galaxy over 1000 light years in diameter. Located 65 million light years from Earth, the prominent dust is a clear indicator of intensive star formation.*

Behind nature, throughout nature, spirit is present; one, and not compound, it does not act upon us from without, that is, in space and time, but spiritually, or though ourselves: therefore, that spirit, that is, the Supreme Being, does not build up nature around us, but puts it forth through us.

RALPH WALDO EMERSON

Whirlpool Galaxy—*Nestled within the spiral arms and dust clouds of the Whirlpool galaxy are young stars, their formation triggered by the gravitational pull of a nearby companion galaxy, NGC 5195.*

Matter is not simply inert or dead, but a numinous reality with both a physical and spiritual dimension. As self-reflective creatures, we need to realize our responsibility for the continuation of the ancient and awesome evolutionary process.

FATHER THOMAS BERRY

Fiddlehead fern spiral.

Kilauea Volcano, Hawaii

God's love doesn't seek value, it creates value.

It is not because we have value that we are loved,

but because we are loved that we have value.

 WILLIAM SLOANE COFFIN

The Cone Nebula (NGC 264)— *resembles a sea monster rising from the deep, but is actually a harmless pillar of gas and dust. Twenty-five hundred light years away, and located in the constellation Monocerus (the Unicorn), the Cone Nebula is believed to be an incubator for developing stars. Its height equals 23 million roundtrips to the Moon.*

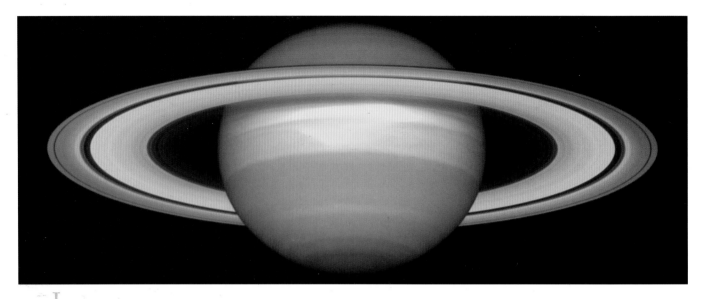

I believe God is the ultimate reality. God is eternal, beyond space and time. The destiny of the human person is to be one with the divine, to be taken up into the life of God.

In my youth I became an agnostic. But I was terribly impressed, as I did research, that the universe really was intelligible. Why does nature always turn out to be more intellectually coherent than anything we can conceive before we do the studies? Why should there be a universe at all? I believe the universe is rational because there is a suprarational Being behind it. I am thrilled by the beauty and rationality of the universe, from quarks to the human brain, its order, intricacy and integration. Personal relationships are a part of that order. They are a clue to the nature of ultimate reality. The personal is the highest category we know, and it can't be reduced to atoms and molecules. It is a reality in its own right. That's why it's justified to conceive of a personal God, because when we do so we are using the language of the highest kind of reality of which we have any experience.

 ARTHUR PEACOCKE

I was sitting through the sermon now every week and finding that I could not only bear the Jesus talk but was interested, searching for clues. I was more and more comfortable with the radical message of peace and equality, with the God in whom Dr. King believed. I had no big theological thoughts but had discovered that if I said, Hello?, to God, I could feel God say, Hello, back. It was like being in a relationship with Casper. Sometimes I wadded up a Kleenex and held it tightly in one fist so that it felt like I was walking hand and hand with him.

 ANNE LAMOTT

Saturn's 30-foot-wide rings are composed of boulder-sized and smaller chunks of ice which, if not disrupted by the planets gravitational field, would otherwise merge and form a moon. The pale reddish color of the rings is due to organic matter mixed with the ice.

North Polar Mosaic— *Earth's moon, as seen from the Galileo spacecraft on December 7, 1992.*

The most detailed true-color image of the entire Earth to date, taken from onboard the Terra satellite (eastern hemisphere).

When you walk on the Earth, imagine that this is the skin of a live being.
Imagine the flow of waters in streams and rivers as the blood of this being.
Imagine the air you breathe as her breath.

WILL BRINTON

Earth (western hemisphere).

The universe is a communion of subjects, not a collection of objects.
And listen to this: The human is derivative. The planet is primary.

FATHER THOMAS BERRY

It would be excessive boldness for anyone to limit and restrict the Divine power and wisdom to some particular fancy of his own.

GALILEO GALILEI

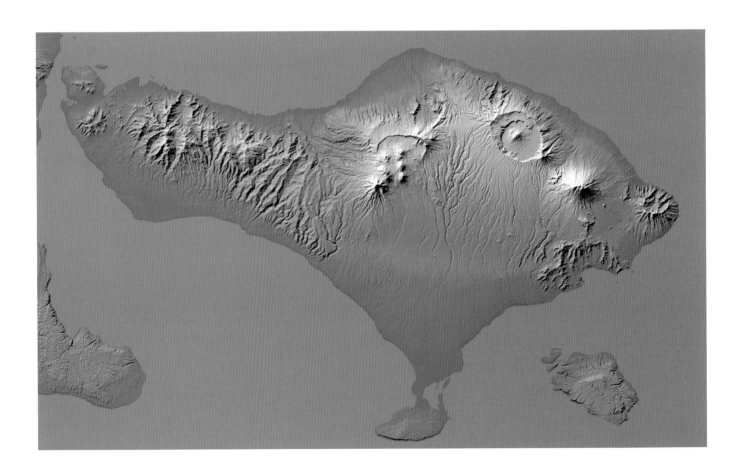

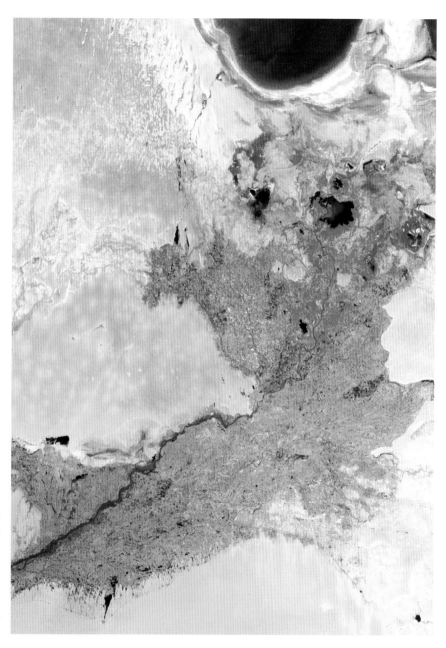

He has made everything beautiful in its time: also he has put eternity into man's mind, yet so that he cannot find out what God has done from the beginning to the end.

ECCLESIASTES 3:11

A relief image of Bali showing the Gunung Agung, a volcano held sacred in Balinese culture, visible at the right center. Inactive for 120 years, the eruption of Gunung Agung in 1963 coincided with a purification ceremony held only once per century.

The Amu Darya River forms a wide delta in the western deserts of Uzbekistan and northeastern Turkmenistan. NASA's SpectralRadiometer acquired this false-color image, where highly vegetated areas appear red.

The majesty of God in itself goes beyond the capacity of human understanding and cannot be comprehended by it ... We must adore its loftiness rather than investigate it, so that we do not remain overwhelmed by so great a splendor.

 JOHN CALVIN

When cloud formations over the ocean are disturbed by wind passing over land, a vortex sheet occurs. Here, marine stratocumulus clouds have arranged themselves in sheets downwind of the Canary Islands.

A dust storm blowing out from Iceland's southern coast.

Only nature has a right to grieve perpetually, for she only is innocent.
Soon the ice will melt, and the blackbirds sing along the river which [my brother]
frequented, as pleasantly as ever. The same everlasting serenity will appear
in this face of God, and we will not be sorrowful, if he is not.

HENRY DAVID THOREAU

I know that Qameta (God) is all-seeing and never sleeps, but I have a suspicion that Qameta may in fact be dozing. If this is the case, the sooner I die the better, because then I can meet him and shake him awake and tell him that the children of Ngubengcuku, the flower of the Xhosa nation, are dying.

 NELSON MANDELA

Overleaf—In this satellite image combined with an elevation model, visible and infrared light reveal details of southeast Alaska's Malaspina Glacier, which is up to 40 miles wide and extends 28 miles from the mountain front to the sea.

An aerial image of Typhoon Sinlaku taken from NASA's Aqua spacecraft, gathering force off the coast south of Japan in September, 2002.

There is a pleasure in the pathless woods,
There is a rapture on the lonely shore,
There is society where none intrudes
By the deep sea, and music in its roar.
I love not man the less, but nature more,
From these our interviews in which I steal
From all I may be, or have been before,
To mingle with the universe, and feel
What I can ne'er express, yet cannot all conceal.

LORD BYRON

For the first time I seriously began to wonder if there is existence after death. It doesn't seem so unlikely as it once did—perhaps because death itself is less remote. I find myself wondering: What if there really is a God? Wouldn't that be extraordinary?

Still, a death sentence does set you apart from the other mortals who will keep on living. You find yourself thinking, I don't have a plan for that, I will not be here. I even feel somewhat removed from my children. But closer to God. I don't know what or who He is, but I am almost sure He is there. I feel His presence, feel that He is close to me during the awful moments. And I feel love. I sometimes feel wrapped, cocooned in love. I often feel it most strongly right before I go to sleep. Then I think of my parents who died years ago and remember what the priest told me when I grieved for my father. "People die," he told me. "They rot and turn to dust. But love is forever."

AGNES COLLARD

Feather worm, San Salvador, Bahamas

Nature is the art of God.

SIR THOMAS BROWNE

I think it pisses God off if you walk by the color purple in a field somewhere
and don't notice it. . . People think pleasing God is all God care about.
But any fool living in the world can see it always trying to please us back.

ALICE WALKER
The Color Purple

Clownfish in a closing Anemone, Solomon Islands

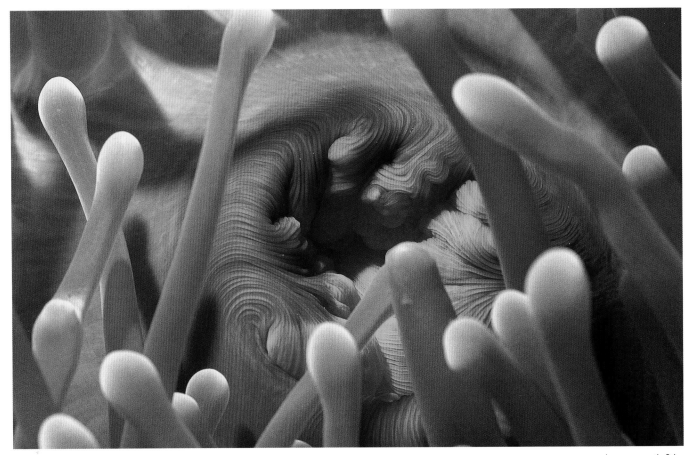

Anemone mouth, Palau

God is not in the vastness of greatness. He is hid in the vastness of smallness.
He is not in the general. He is in the particular.

 PEARL S. BUCK

Julian of Norwich talks of putting on God like a garment. It is a homely and appealing picture of comfort and safety. I imagine a favorite roomy sweater, a little baggy in just the right places, or maybe a soft old bathrobe.

MARGARET GUENTHER

God is in the prepositions–beyond, among, within, beneath.

SHARON DALOZ PARKS

Soft coral, Palau

If thy heart were right, then every creature would be a mirror of life and a book of holy doctrine. There is no creature so small and abject, but it reflects the goodness of God.

THOMAS À KEMPIS

Nudibranch (Chromodoris kuniei), Solomon Islands. There are over 3000 varieites of nudibranch, also known as the sea slug, a shell-less snail that originated in Europe.

Hermit Crab, Bonaire

And the earth was without form and void, and darkness was upon the face of the deep;
And the Spirit of God moved upon the face of the waters.

GENESIS 1: 2

Nudibranch (Nembrotha Kubaryana), Solomon Islands

God is a pure no-thing,
concealed in now and here;
the less you reach for him,
the more he will appear.

 ANGELUS SILESIUS

There is a twofold meaning in
every creature, a literal and a
mystical, and the one is but the
ground of the other.

JOHN SMITH

Nudibranch (Notodoris minor), Solomon Islands

There is grandeur in this view of life, with its several powers, having been originally breathed by the Creator into a few forms or into one; and that, whilst this planet has gone cycling on according to the fixed law of gravity, from so simple a beginning endless forms most beautiful and most wonderful have been, and are being, evolved.

CHARLES DARWIN

More appropriate, I should think, is the view that God created the universe out of an interest in spontaneous creativity—that he wanted nature to produce surprises, phenomena that he himself could not have foreseen. What would such a creative universe be like? Well, it would for one thing be impossible to predict in detail. And this seems to be the case with the universe we inhabit. . . Further, a creative universe should give rise to agencies that are themselves creative, which is to say unpredictable. There is in our universe such an agency, spectacularly successful at reversing the dreary slide of entropy and making surprising things happen. We call it life.

TIMOTHY FERRIS

Cuttlefish, Solomon Islands

83

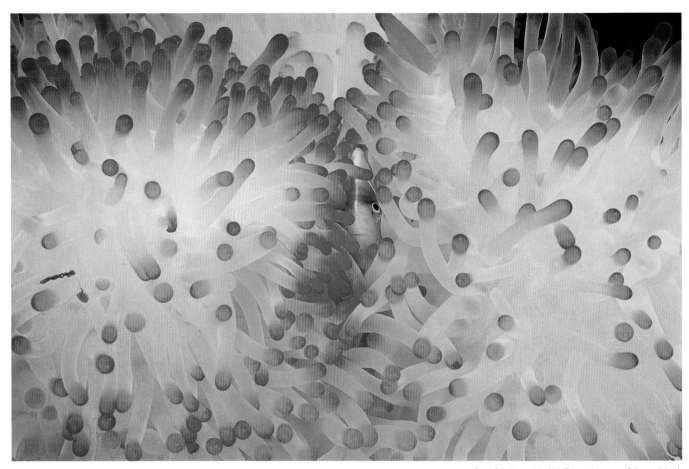

Clownfish (center) in Pink-Tipped Anemone, Solomon Islands

Nature is full of genius, full of the divinity; so that not a snowflake escapes its fashioning hand.

HENRY DAVID THOREAU

Humans need elements of the natural to make and keep life human. A society attuned to artifacts forgets creation; maybe that's New York versus the Rocky Mountains. What does it profit a man to gain the world only to lose it? To consume the world and lose soul in the tradeoff? Nature invites us to think of our sources, of the Great Source, more than of resources. The most authentic wilderness emotion is the sense of the sublime. We get transported by forces awe-full and overpowering, by the signature of time and eternity.

HOLMES ROLSTON, III

Scorpionfish, Curacao

Peacock Flounder, San Salvador, Bahamas

To see a World in a Grain of Sand and a Heaven in a Wild Flower
Hold Infinity in the palm of your hand, and Eternity in an hour.

WILLIAM BLAKE

You ask: What is the meaning or purpose of life? I can only answer with another question:
Do you think we are wise enough to read God's mind?

FREEMAN DYSON

Lionfish, Solomon Islands

Earth is a kind of providing ground, where the life epic is lived on in the midst of its perpetual perishing. Life persists because it is provided for in the evolutionary and ecological Earth systems. Today we say: life is generated "at the edge of chaos." Yesterday, John said: "The light shines in the darkness, and the darkness has not overcome it" (John 1: 5). I think the Twenty-Third Psalm is pretty good experiential biology: Life is lived in green pastures and in the valley of the shadow of death, nourished by eating at a table prepared in the midst of its enemies.

HOLMES ROLSTON, III

Reef Shark, Walkers Cay, Bahamas

believe that God has many attributes, which He shows at different times. In Hebrew there are about 60 names for God. Each one denotes a characteristic. When God is using judgment more, we say one name. When God is being merciful more, we say another. He can be angry, and He can be kind. But they're not extremes; they're one in the same.

I don't see Him as moody, but He is emotional, and He wants an emotional relationship with man. He's creative—of course He is! Look at the world He has made. Everything in it is unique.

He's witty. Not sarcastic, but witty. And sharp, very, very sharp. If He were to say something, it would be a one-liner. Just perfect.

CHANA MEIER

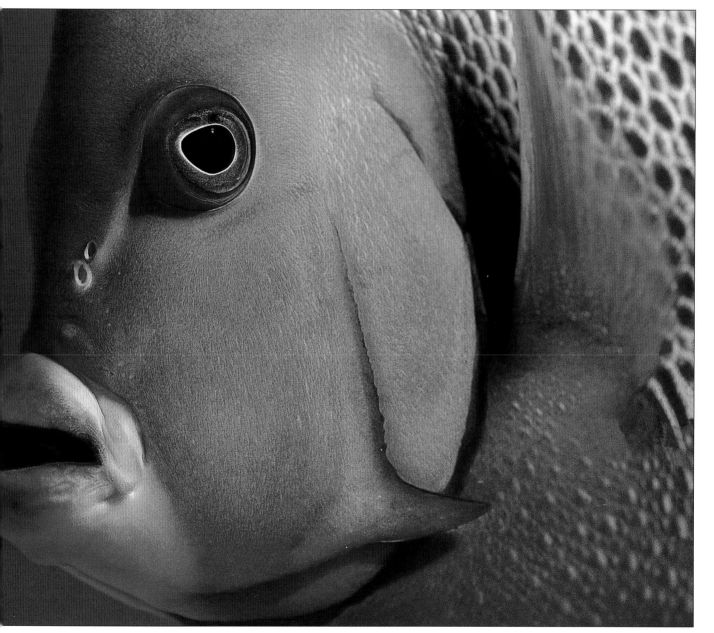

Gray Angelfish, Cayman Islands

I went through many stages.
At times I felt that God was cruel,
that God was absent.
The main thing I felt was that
God was silent. But it's still a
question: Was he cruel? Was He
trying to punish for love?
My questioning of God goes on.
But even from the beginning I
believed in questioning God from
inside faith, not outside faith.
It is because I believe that I am
all the time questioning.

ELIE WIESEL

It is the sense of mystery that, in my opinion, drives the true scientist; the same blind force, blindly seeing, deafly hearing, unconsciously remembering, that drives the larva into the butterfly. If [the scientist] has not experienced, at least a few times in his life, this cold shudder down his spine, this confrontation with an immense invisible face whose breath moves him to tears, he is not a scientist.

ERWIN CHARGAFF

Adult and juvenile humpback whales

In the eighth chapter of Romans, Paul gives us a conception of nature as anything but complete and perfect and with it a conception of God as anything but the remote absentee landlord. Not all in the garden is lovely; neither in the garden of nature nor in the life of man. But that is precisely the sort of universe in which God can be involved. "Up to the present we know the whole created universe groans in all its parts as if in pangs of childbirth. Not only so, but even we . . . are groaning inwardly" (Romans 8: 22-23). Here is recognition of struggle and cruelty and pain in nature. But it is a struggle pregnant with possibility, the possibility of new birth. It is a struggle with a hope in it. And as we read on, we see the great synthesis of Paul's imaginative thought leads him to see a parallel between nature's struggle and our own. It is a picture of unfinished man in an unfinished universe. Somehow or other both nature and man are incomplete. estranged and separated from what they could be and eventually might be.

The word creation is one of the great symbol-words describing
the relation of God to the Universe.

PAUL TILLICH

A brilliant, fountainlike eruption spills a river of molten lava
from a volcano in the Ring of Fire, a region on the edge of
the Pacific Ocean.

Panoramic view of a sunset over Island River near Lake Superior.

Fiery waterfall from Hawaii's active Kilauea volcano pours down a cliff and into the ocean.

I worship Lord Sankara. I picture Him in my mind as His idol in the temple. I pray that I get relief from this horrible disease. That's all I ask. I have not committed any sins in this life. I was an ordinary laborer. Then four years ago I was stricken with leprosy and my life changed. Brahma has written out my fate. I am being punished for sins I must have committed in my last life. God is vengeful. He really punishes. There is no escape from the consequences of sin. I pray that heaven will be better than this.

 BALDEVA RAM BENARES

Summer storm in Yellowstone National Park.

God is like a mirror. The mirror never changes, but everybody who looks at it sees something different.

RABBI HAROLD KUSHNER

An aerial view of the Grand Prismatic Spring in Yellowstone.

Yellowstone's Geyser Basin at twilight.

I find it interesting that the meanest life, the poorest existence, is attributed to God's will, but as human beings become more affluent, as their living standard and style begin to ascend the material scale, God descends the scale of responsibility at a commensurate speed.

MAYA ANGELOU

A rain forest seen from above in Borneo's Gunung Palung National Park, West Kalimantan.

I sat there and forgot and forgot, until what remained was the river that went by and I who watched. . . Eventually the watcher joined the river and then there was only one of us. I believe it was the river.

NORMAN MACLEAN

If the detail of the happenings in the universe were all determined in exactness, then there could be no creation. God sustains and lures, but there is spontaneity of created nature to respond or not to respond. That is the element of freedom of the creation. To be consistent, we must provide for freedom at the most elementary level of matter. Without that the idea of creation disappears. God's creativity involves the spontaneity and freedom of the creature. This is an idea which is at the base of the Christian concept of the nature of man. But is this not also a way of looking at the whole creation? The purposes of God in creation are not implemented as a series of arbitrary acts, but as a struggle between a disordered state and God's lure to completeness.

L. CHARLES BIRCH

The setting sun blazes across the face of granite El Capitan and illuminates the Merced River below in Yosemite National Park.

The intuitive mind is a sacred gift and the rational mind is a faithful servant. We have created a society that honors the servant and has forgotten the gift.

ALBERT EINSTEIN

Millaa Millaa Falls, Queensland, Australia

God is not what you imagine or what you think you understand.
If you understand you have failed.

 SAINT AUGUSTINE

Old-growth fir trees in one of Germany's national forests.

What I know of the Devine
I learnt in the woods and fields.
I have no other masters
Other than the beeches and the oaks.

SAINT BERNARD

The redwoods, once seen, leave a mark or create a vision that stays with you always. No one has ever successfully painted or photographed a redwood tree. The feeling they produce is not transferable. From them comes silence and awe. It's not only their unbelievable stature, nor the color which seems to shift and vary under your eyes, no, they are not like any trees we know, they are ambassadors from another time.

JOHN STEINBECK

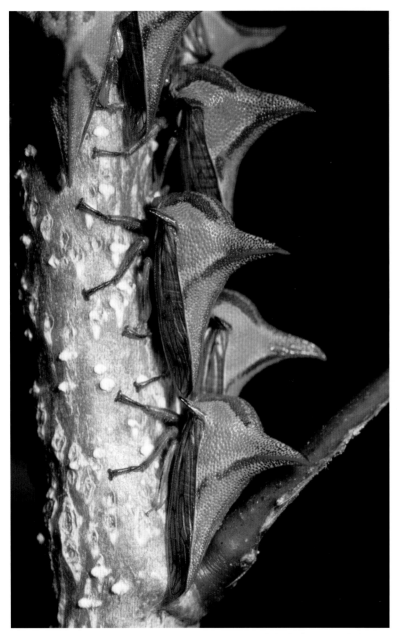

When one realizes oneself, one realizes the essential nature of the universe. The existence of duality is only an illusion and when the illusion is undone, the primordial unity of one's own nature and the nature of the universe is realized, or made real.

 NAMKHAI NORBU

Thorn-mimic treehoppers campylenchia latipes—perched on twig.

The "praying posture" or threat display, demonstrated by a rain forest mantid in Gunung Palung National Park, Indonesia.

I have never understood why it should be considered derogatory to the Creator to suppose that he has a sense of humour.

WILLIAM RALPH INGE

There is a story I once heard about a young boy who went out to the woods day after day. His father took note of this strange habit and asked his child, "My son, why do you go out to the woods each day?" The son responded, "I go there to find God." At this the father gently reprimanded his child: "Don't you know that God is the same everywhere?" The son replied, "Yes, Father, but I am not the same everywhere."

UNKNOWN

Dragonfly

American columbine

We are in danger of forgetting that we cannot do what God does, and that God will not do what we can do.

OSWALD CHAMBERS

The supernatural is the natural not yet understood.

ELBERT HUBBARD

The more one studies the Christian conception of the Divine, the more subtle and complex it is found to be. There is something impenetrable about it. It is as though one comes to a frontier of the knowable; beyond lies that which, to the human mind, will never be completely understood. This is how it should be. God, as we must expect, is all that we are and more besides. By "more" I mean not only bigger, better, or more powerful, but also "more" in the sense of being different—different in some fundamental, subtle, and complex way that borders on the unimaginable, the unthinkable.

 RUSSELL STANNARD

The Puss Moth (Cerura vinula) caterpillar, showing its false face.

I say to mankind, Be not curious about God. For I, who am curious about each,
am not curious about God—I hear and behold God in every object,
yet understand God not in the least.

WALT WHITMAN

Like the reflections of the moon that effortlessly appear in any body of still water, a Buddha's emanations spontaneously appear wherever living beings' minds are capable of perceiving them. Buddhas can emanate in any form whatsoever to help living beings. Sometimes they manifest as Buddhists and sometimes as non-Buddhists. They can manifest as women or men, monarchs or tramps, law-abiding citizens or criminals. They can even manifest as animals, as wind or rain, or as mountains or islands. Unless we are a Buddha ourself we cannot possibly say who or what is an emanation of a Buddha.

GESHE KELSANG GYATSO

Stunning sapphire-hued Dendrobates azureus, or poison dart frogs. Their poison is used by some South American Indians for applying to the tips of their hunting arrows and blow-gun darts.

Karst limestone tower in Vietnam's Ha Long ("Descending Dragon") Bay, an area of great beauty as well as spiritual importance located in the South China Sea. Some 3000 limestone and dolomite islands rise jaggedly from the Gulf of Tonkin like the scales on a dragon's back.

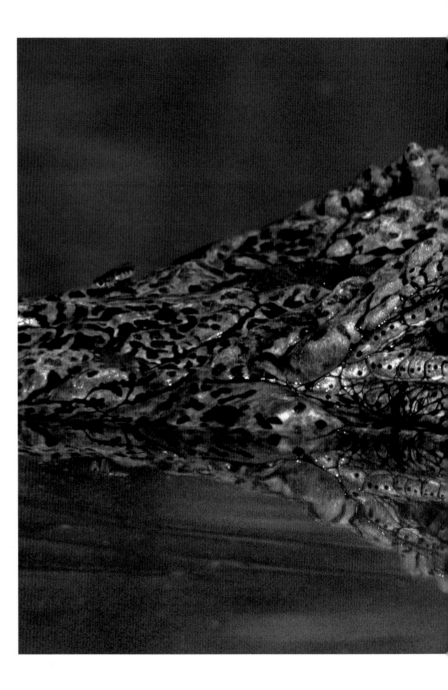

I cannot think that the world as we see it is the result of chance; and yet I cannot look at each separate thing as the result of design... I am, and shall ever remain, in a hopeless muddle.

If anything is designed, certainly man must be, yet I cannot admit that man's rudimentary mammae were designed... I am in a thick mud yet I cannot keep out of the question.

CHARLES DARWIN

Crocodile on the Nile, Kenya, Africa.

My atheism, like that of Spinoza, is true piety towards the universe and denies only gods fashioned by men in their own image to be servants of their human interests.

GEORGE SANTAYANA

Question with boldness even the existence of a God; because, if there be one, he must more approve of the homage of reason, than that of blind-folded fear.

THOMAS JEFFERSON

God must be allowed the right to speak unpredictably.

THOMAS MERTON

That God of the clergymen, he is for me as dead as a doornail. But am I an atheist for all that? The clergymen consider me as such—be it so; but I love, and how could I feel love if I did not live, and if others did not live, and then, if we live, there is something mysterious in that. Now call that God, or human nature or whatever you like, but there is something which I cannot define systematically, though it is very much alive and very real, and see, that is God, or as good as God. To believe in God for me is to feel that there is a God, not a dead one, or a stuffed one, but a living one, who with irresistible force urges us toward "aimer encore"; that is my opinion.

VINCENT VAN GOGH

The three-horned chameleon exists only on the Congo (Zaire) side of the Virunga volcanoes. It figures prominently in African folklore and mythology, and is often linked with errands of great importance and issues of mortality.

A red-eyed tree frog clings to a leaf in Costa Rica.

We can see infinity and eternity in the variety of facial features, temperaments, and affections of people. We can also see among the many types of plants, animals, and humans that the drive to propagate is part of a never-ending endeavor to reproduce—from creation to eternity. We can also see this in the many areas that people have explored wisdom and the realization that we can never know everything. There will always be areas to explore and new knowledge to gain. The fact that these expressions of eternity and infinity exist within human expression of reproduction and the search for wisdom is evidence of the influence of God, who is infinity and eternity. However, since humans are finite and can have nothing of the infinity within them except as emanations of God, they can be seen merely as the vessels that receive this love and wisdom, also known as infinity and eternity, even though it appears as though they act from these qualities from their own natures.

JOANNA V. HILL
Divine Providence, according
to the Writings of Emanuel Swedenborg

The Ulysses Butterfly is also known as the Blue Mountain Swallowtail, the Blue Emperor, and the Mountain Blue. A spectacular, bright blue and black Australian swallowtail butterfly, it lives in tropical rain forests in Australia, Indonesia, and other nearby islands.

A flap-necked chameleon hugs a leaf bud in Tanzania, Africa.

But what sort is this God? First he maliciously refused Adam from eating of the tree of knowledge, and, secondly, he said "Adam, where are you?" God does not have foreknowledge? Would he not know from the beginning? And afterwards, he said, "Let us cast him out of this place, lest he eat of the tree of life and live forever." Surely, he has shown himself to be a malicious grudger! And what kind of God is this? For great is the blindness of those who read, and they did not know him. And he said, "I am the jealous God; I will bring the sins of the fathers upon the children until three (and) four generations." And he said, "I will make their heart thick, and I will cause their mind to become blind, that they might not know nor comprehend the things that are said." But these things he has said to those who believe in him and serve him!

GNOSTIC TREATISE
The Testimony of Truth

A king cobra swallows another snake.

Since God is substance itself and form itself, the only and thus the first, whose essence is love and wisdom, and since from Him all things were made which are made, it follows that He created the universe with everything in it from Love by means of Wisdom; and consequently that the Divine Love, together with the Divine Wisdom, is in every created subject. Love, moreover, not only is the essence which forms all things, but it also unites and conjoins them, and thus keeps them in connection when formed.

EMANUEL SWEDENBORG

Human subtlety will never devise an invention more beautiful, more simple
or more direct than does nature because in her inventions nothing
is lacking, and nothing is superfluous.

LEONARDO DA VINCI

God is the great mysterious motivator of what we call nature, and it has often been said by philosophers, that nature is the will of God. And I prefer to say that nature is the only body of God that we shall ever see.

FRANK LLOYD WRIGHT

I love to think of nature as an unlimited broadcasting station, through which God speaks to us every hour, if we will only tune in.

GEORGE WASHINGTON CARVER

One of the largest new world vultures, with a wing span of four feet, the King Vulture inhabits the dense tropical lowland forests of Central America, from southern Mexico to Argentina.

Grey-crowned crane, Africa.

It's a popular fact that 90% of the brain is not used and, like most popular facts, it is wrong. Not even the most stupid Creator would go to the trouble of making the human head carry around several pounds of unnecessary grey goo if its only real purpose was, to serve as a delicacy for certain remote tribesmen in unexplored valleys, it is used. One of its functions is to make the miraculous seem ordinary, and turn the unusual into the usual. Otherwise, human beings, forced with the daily wondrousness of everything, would go around wearing a stupid grin, saying "WOW" a lot. Part of the brain exists to stop this happening. It is very efficient, and can make people experience boredom in the middle of marvels.

TERRY PRATCHETT

The American bald eagle (haliaeetus leucocephalus). Eagles once numbered around 50,000 in the contiguous United States, but by 1972, only about 800 breeding pairs remained. Under the Endangered Species Act of 1973, however, the eagles have made a steady recovery. Breeding pairs now number close to 3000; only in Canada and Alaska, however, are eagles found in abundance.

The Jews seek after miracles, they are beset with an idea of God as a God of power, a Sultan, a miracle worker. The Greeks are crazy for wisdom, they think of God in terms of the supreme Philosopher, the supreme Mathematician, the supreme Engineer. But we are not satisfied with a picture of God in terms either of power or wisdom. We see God in terms of a man on a cross, of the love that suffers, and suffering redeems. We see God not in terms of a Sultan, nor an Engineer, neither of Judge nor of Designer. We see God in terms of our Father.

F. E. RAVEN

Maligne Lake, located in Jasper National Park in Alberta, Canada.

Bison herd in Hayden Valley, Yellowstone National Park.

The forest is not merely an expression or representation of sacredness nor a place to invoke the sacred–the forest is sacredness itself. Nature is not merely created by God, nature is God. Whoever moves within the forest can partake directly of sacredness, experience sacredness with his entire body, breath sacredness and contain it within himself, drink the sacred water as a living communion, bury his feet in sacredness, open and witness the burning beauty of sacredness.

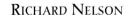

RICHARD NELSON

It may be that our role on this planet is not to worship God, but to create him.

Grizzly at Hallo Bay, Alaska

Silhouetted acacia tree

I believe God is everything . . . Everything that is or ever was or ever will be. And when you can feel that, and be happy to feel that, you've found It. . . . My first step from the old white man was trees. Then air. Then birds. Then other people. But one day when I was sitting quiet and feeling like a motherless child, which I was, it come to me: that feeling of being part of everything, not separate at all. I knew that if I cut a tree, my arm would bleed. And I laughed and I cried and I run all round the house. I knew just what it was. In fact, when it happen, you can't miss it.

ALICE WALKER
The Color Purple

Herd of Cape buffalo

The Christian God is pre-eminently a God of love. Love of its very nature must reach out to someone else. Once it turns in on itself and becomes self-love it degenerates into selfishness and self-indulgence—the attributes that are the opposite of love. For the greater part of time, the universe had no living creatures. During that stage of development, had God been a single entity, there would have been no love, for there would have been no one for God to love except himself. So, from the beginning, the existence of love required there to be a relationship of some kind. The overall unity of God had to embrace an inner self-relatedness. Creation, therefore, marked not so much the start of God's relationships, as the outward expression of the inner structure of God's own being.

Russell Stannard

The worldview that modern science has depended on, for the most part, is that if you have 100 percent knowledge of an initial condition, then all subsequent conditions are known as well. If this were true, there would be no such thing as free will, whether human or divine. God himself would be a merely passive observer who not only doesn't but *couldn't* have an effect on the world. And human beings couldn't have an effect either, because if all physical events are, in effect, laid out ahead of time, like a complex play in billiards, then nothing one does can alter the outcome. It's all completely determined. This is the fundamental premise of the modern worldview. However, it turns out and what modern physics has now demonstrated is, that's not true. That actually, at the most fundamental level, all the most important physical processes are, in part, determined by "factors" that have no detectable presence in the physical world. A range of possible outcomes are determined mechanically, but untold numbers of decisions are being made by "something" that from among these possibilities selects every actual outcome. And furthermore, each time a decision is made, all other probabilities are instantaneously adjusted and altered so as to keep the whole system within certain bounds. This is not a philosophical concept. This is a description of what has been shown, to the shock and horror of many scientists, in actual physical experiments. The only way to talk about it metaphorically is that there's "something" that is not part of the physical universe, which sits outside it, and simultaneously orchestrates all events throughout the entire universe, according to principles that we can't know. My own particular angle on this is that if there is this effect that is sustaining the universe in an ongoing way, causing results, but it itself has no prior cause, well, that is one of the oldest definitions of God.

 JEFFREY B. SATINOVER M.D.

Cheetah stalking a herd of wildebeest

Harmony is only in following the Way.

The Way is without form or quality,
But expresses all forms and qualities;
The Way is hidden and implicate,
But expresses all of nature;
The Way is unchanging,
But expresses all motion.

Beneath sensation and memory
The Way is the source of all the world.
How can I understand the source
of the world?
By accepting.
LAO TZE

When you are in the presence of the Holy Spirit, it is like sittting in front of a fire that does not burn you, but suffuses you with its qualities—its warmth, glow, and color. And as you are there, in the presence of the Spirit, you also become suffused with the divine attributes of compassion, gentleness, and love, without your doing anything about it except to be there. You are loved and you are held in this love.

DESMOND TUTU

Faith is the word that describes the direction our feet start moving when we find that we are loved. Faith is stepping out into the unknown with nothing to guide us but a hand just beyond our grasp.

 FREDERICK BUECHNER

Is it your thought that I despise some of these, while I love the others? I tell you, I despise nothing. None of it is repulsive to Me. It is life, and life is the gift; the unspeakable treasure; the holy of holies.

NEALE DONALD WALSCH
Conversations with God

Sleeping warthog

The knowledge of the soul admittedly contributes greatly to the advance of truth in general, and, above all, to our understanding of Nature, for the soul is in some sense the principle of animal life.

ARISTOTLE

ake any of our five senses out
of our bodies and we could still
survive. Take our breath out of
our bodies and we could not.
We have this in common with
everyone who is alive; we all
live on the air we breathe. The
image that we generally have of
God is that He is somewhere in
the air, somewhere above us in
heaven. But air is above us,
around us, and inside our lungs.
As the author Agnes Sanford
has observed, "We live in God.
That's what we breathe."

LEONTYNE PRICE

Furthermore, it's important to remember that God not only made our body and our mind (through the natural processes He created), but He also gave us our soul or spirit, which is the part of us that is able to know God. We aren't just bodies or minds; if we were, we would be like every other animal (although unique because of our abilities). We also were created in the image of God.

REV. BILLY GRAHAM

All are but parts of one stupendous whole,
Whose body Nature is, and God the soul.

ALEXANDER POPE

He stared out at the ocean and said, "Look at the view, young lady. Look at the view." And every day, in some little way, I try to do what he said. I try to look at the view. That's all. Words of wisdom from a man with not a dime in his pocket, no place to go, nowhere to be. Look at the view. When I do what he said, I am never disappointed.

ANNA QUINDLEN

Mandrill in captivity

God gives what He has, not
what he has not: He gives the
happiness that there is, not the
happiness that there is not.
To be God—to be like God and
to share His goodness in creaturely
response—to be miserable—these
are the only three alternatives.
If we will not learn to eat the only
food that the universe grows—the
only food that any possible
universe ever can grow—then
we must starve eternally.

C. S. LEWIS

God is constantly creating, in us, through us, and with us,
and to co-create with God is our human calling.

MADELEINE L'ENGLE

Sometimes I wake up in the morning and I can just hear melodies and little themes, and I know that it is directly from God because it is so pure, it's good, it just came through me. It doesn't happen with me a lot. With someone like Louis Armstrong–all the time. It depends on the musician. It depends on God.

WYNTON MARSALIS

Asian elephant's eye

Tiger got to hunt, bird got to fly;
Man got to sit and wonder "why, why, why?"
Tiger got to sleep, bird got to land;
Man got to tell himself he understand.

KURT VONNEGUT
Cat's Cradle

All species of tigers are endangered, and of the world's eight
Panthera tigris subspecies, three are extinct—the Caspian, the
Bali and the Javan. Remaining are the Bengal, South China,
Indochinese, Sumatran and Siberian. A commonly cited estimate
of the number of living tigers today is between 5,000 and
7,000. To survive, they will require massive human intervention.

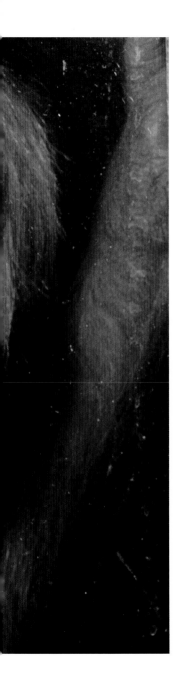

I tell you this: You always get what you create, and you are always creating. I do not make a judgment about the creations that you conjure, I simply empower you to conjure more—and more and more and more. If you don't like what you've just created, choose again. My job, as God, is to always give you that opportunity.

I truly want what you truly want—nothing different and nothing more. Don't you see that is My greatest gift to you? If I wanted for you something other than what you want for you, and then went so far as to cause you to have it, where is your free choice? How can you be a creative being if I am dictating what you shall be, do, and have? My joy is in your freedom, not your compliance.

I am telling you that your perception of ultimate reality is more limited than you thought, and that Truth is more unlimited than you can imagine.

Forever is longer than you know. Eternal is longer than Forever. God is more than you imagine. God is the energy you call imagination. God is creation. God is first thought. And God is last experience. And God is everything in between.

NEALE DONALD WALSCH
Conversations with God

This is my father's world, and to my listening ears.
All nature sings, and round me rings.
The music of the spheres.
This is my father's world; I rest me in the thought.
Of rocks and trees, of skies and seas;
His hand the wonders wrought.

This is my father's world; the birds their carols raise;
The morning light, the lily white,
Declare their maker's praise.
This is my father's world; he shines in all that's fair.
In the rustling grass I hear him pass;
He speaks to me everywhere.

This is my father's world; oh, let me ne'er forget
That, though the wrong seems oft so strong,
God is the ruler yet.
This is my Father's world: why should my heart be sad?
The Lord is King; let the heavens ring!
God reigns; let the earth be glad!

MALTBIE D. BABCOCK

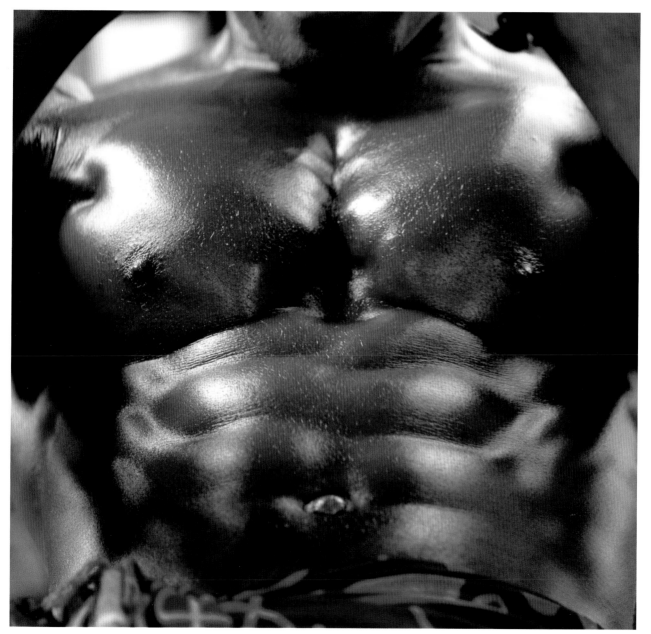

Upper body, weightlifter

This world is indeed a living being endowed with a soul and intelligence ... a single visible living entity containing all the other living entities, which by their nature are all related.

PLATO

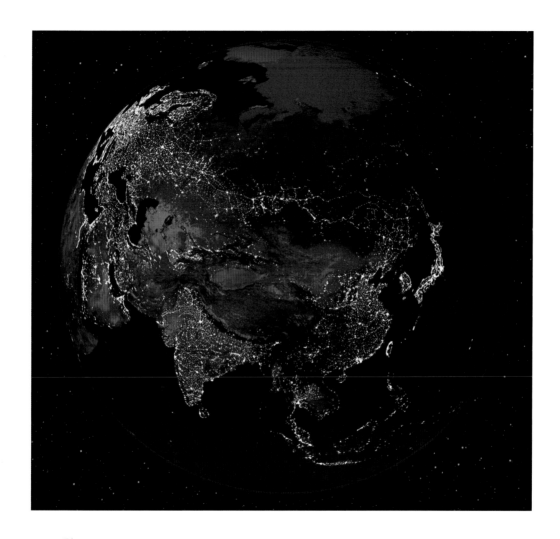

Someday, after mastering the winds, the waves, the tides and gravity, we will harness for God the energies of love. And then, for the second time in the history of the world, man will discover fire.

TEILHARD DE CHARDIN

The more I know, the less I understand
All the things I thought I knew I'm learning again
I've been trying to get down to the heart of the matter
But my will gets weak and my thoughts seem to scatter
But I think it's about forgiveness, forgiveness.

DON HENLEY

Reflections on the Nature of God—Photo Credits

NATIONAL GEOGRAPHIC IMAGE COLLECTION

Jonathan Blair: 145,
Skip Brown: 134-135
Jodi Cobb: 149, 155
Nicole Duplaix: 121
John Eastcott and Yva Momatiuk: 127, 136-137, 138
Jason Edwards: 13, 124
Raymond K. Gehman: 96-97, 99, 100, 128
George Gral: 12, 110, 114
Roy Gumpel: 158
Bill Hatcher: 115
Wolcott Henry: 92-93
Chris Johns: 142, 144
Beverly Joubert: 133, 139, 143
Mattias Klum: 123
Tim Laman: 57, 102, 108, 109, 125, 126, 146
Bates Littlehales: 112
Peter Lik: 104-105
Robert Madden: 94
George F. Mobley: 132
Darlyne A. Murawski: 113
Michael Nichols: 49, 118, 151
Paul Nicklen: 107
Richard Nowitz: 152
Norbert Rosing: 106, 130
Randy Olson: 101
Joel Sartore: 15, 131, 148
Phil Schermeister: 103
Maria Stenzel: 71
Volkmar K. Wentzel: 141
Steve Winter: 119

Todd Gustafson/Panoramic Images/NGSImages.com: 116-117, 122, 140
Manuel Magana: 46, 52, 74, 75, 76, 77, 78, 79, 80, 81, 83, 84, 85, 86, 87, 89, 90-91
Douglas Peebles/Panoramic Images/NGSImages.com: 59, 98
Barbara Youngleson: 72, 111, 147

ACS Science & Engineering Team, NASA: 48
European Space Agency and Wolfram Freudling (Space Telescope-European Coordinating Facility/European Southern Observatory, Germany): 39

Hubble Heritage Team: 4
Hubble Heritage Team, A. Riess (STScI) NASA: 32
Jacques Descloitres, MODIS Land Rapid Response Team, NASA/GSFC: 66, 67, 70
Local Group Galaxies Survey Team/NOAO/AURA/NSF: 37
Marit Jentoft-Nilsen, NASA GSFC Visualization Analysis Lab: 54
M. Neeser (Univ-Sternwarte), P Barthel (Kapteyn Astron Institute), H. Heyer. H. Boffin (ESO): 55
NASA: 20, 21, 33, 34, 47, 61, 62, 63,156, 157
NASA, ESA, S. Beckwith (STScI) and the HUDF Team: 8
NASA, ESA, and The Hubble Heritage Team (AURA/STScI): 28
NASA/GSFC/METI/ERSDA/JAROS, and U.S./Japan ASTER Science Team: 16
NASA/GSFC/METI/ERSDAC/JAROS, and U.S./Japan ASTER Science Team: 50
NASA/GSFC/LaRC/JPL, MISR Team: 65
NASA/Hubble: 53
NASA, H. Ford (JHU), G. Illingworth (UCSC/LO), M. Clampin (STScI): 58
NASA, JPL: 6
NASA/JPL/Caltech: 41
NASA/JPL/Hubble Heritage Team (STScI/AURA): 36
NASA/JPL/NIMA: 64, 68-69
NASA, Space Telescope Science Institute: 10
NASA and The Hubble Heritage Team (AURA/STScI): 23, 24, 31, 40, 44, 56, 60
NASA, Jayanne English (University of Manitoba), Sally Hunsberger (Pennsylvania State University), Zolt Levay (Space Telescope Science Institute), Sarah Gallagher (Pennsylvania State University), and Jane Charlton: 29
NASA, N. Walborn and J. Máiz-Apellániz (Space Telescope Science Institute). R. Barbá (La Plata Observatory, La Plata, Argentina): 43
Schaller for STScI: 51
N. Smith (U. Colorado), J. Morse (Arizona State U.), and NASA: 19
T. Rector/University of Alaska Anchorage, WIYN and NOAO/AURA/NSF: 27
University of Colorado, University of Hawaii and NOAO/AURA/NSF: 22